Nicolaus Copernicus

AN ESSAY ON HIS LIFE AND WORK

Nicolaus Copernicus

AN ESSAY ON HIS LIFE AND WORK

FRED HOYLE

Heinemann

Heinemann Educational Books Ltd

London Edinburgh Melbourne Auckland Toronto
Hong Kong Singapore Kuala Lumpur
Ibadan Nairobi Johannesburg New Delhi

ISBN 0 435 54425 X

© Fred Hoyle 1973

First published 1973

Published by Heinemann Educational Books Ltd,
48 Charles Street, London W1X 8AH

Printed in Great Britain by
Richard Clay (The Chaucer Press), Ltd,
Bungay, Suffolk

Preface

IT IS UNEXPECTEDLY AWKWARD TO DESCRIBE exactly what Copernicus did, harder than to write an account of the achievements of Kepler, just because it is less difficult to write about the complete answer to a problem than about an approximate one. Kepler's solution to the problem of the planetary motions was essentially complete, while that of Copernicus was approximate. To understand Copernicus we need therefore to consider how good his approximations really were, and this kind of question always turns out to be difficult.

The difficulty is compounded by the style of the work. Mathematicians and astronomers to the time of Newton used intricate geometrical constructions, which today we find unfamiliar. We prefer nowadays to use symbols and equations instead of pictorial representations, because analytical methods have turned out to be much more powerful—results are derived with economy of effort, and are compact and therefore more easily understood. The proof that the planets move in elliptic orbits is a good example. Using modern methods this result can be obtained from only a page of working and is readily comprehended by university students in a first course on mechanics. In contrast, Newton's original proof was probably not understood in his own day by more than a dozen people.

When I first encountered the geometrical respresentations of Copernicus and Ptolemy I wondered how they could be described analytically. The answer turned out to be unexpectedly elegant. It became possible to interpret the many curious features of the geometry which they used. So on this five-hundredth anniversary of the birth of Copernicus I decided to write an account of the technical meaning of his work. This intention is covered in the present book by Chapters I, III, IV, and by the Epilogue.

It appeared worth while to add the biographical sketch of Chapter II. In the outcome this sketch has caused me some difficulty. My first idea was to abstract from standard biographies the aspects of the life of Copernicus which seemed relevant to his astronomical achievements. After consulting such accounts, of which the three volumes by Leopold Prowe (Weidmannsche Buchhandlung, Berlin 1883–4) are the most complete, I found myself unable to answer certain simple and pertinent questions, particularly questions concerning the periods when the actual astronomical discoveries were made. I also found aspects of these accounts which I felt to be implausible, at any rate from the point of view of the working scientist. This inevitably led me into issues of interpretation, which I would much have preferred to avoid. For example, was *Commentariolus* written pre-1512 or post-1530? Above all, I became overwhelmed by the fact, discovered in the nineteenth century, that *de revolutionibus* was not set by the printer from Copernicus' own manuscript. Why? A fair copy must have been made by someone. Who? These questions contain implications which I have discussed towards the end of Chapter II.

It is once again my pleasure to thank Jan Rasmussen and Evaline Gibbs for their very considerable help in the preparation of the manuscript of this book.

1973   F. H.

List of Plates

Contents

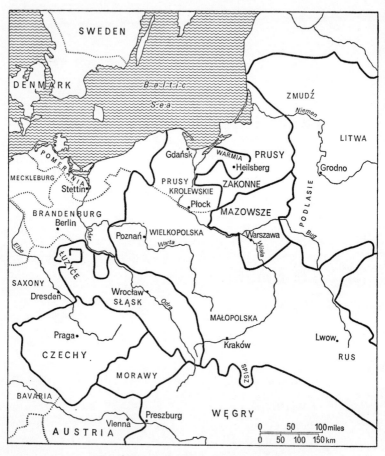

FIGURE 1. North-central Europe showing the province of
Warmia at the time of Copernicus

the modern mind, it was not at all clear in the sixteenth century. Tycho Brahe proposed a dualistic scheme, with the Sun going around the Earth but with all other planets then going around the Sun, and in making this proposal he believed he was offering something radically different from Copernicus. And in rejecting Tycho's scheme, Kepler obviously thought so too. Yet in principle there is no difference.

So what was the issue? The issue was to obtain even *one* substantially correct empirical description of the planetary motions. The issue was to find out *how* the planets moved. Not until this problem was solved could the more far-reaching problem of finding out *why* be tackled effectively. The essential facts had to come before the dynamical theory could be developed. The modern student learns the answer to the problem of *how* in one quick gulp. Each planet moves essentially in an ellipse with the Sun at one of the foci. To describe the actual motions one needs to know the orbital period of each planet, the size and eccentricity of its ellipse and the orientation, and the position of each planet at some instant of time. With hindsight, knowing the answer, the situation may not seem unduly complicated, but looked at without foreknowledge the problem of *how* is anything but simple. This can be seen from Plate 1, which simulates the bare facts of how a terrestrial observer sees the planets move over a period of time. Starting with observations of this kind, discovering the rule of elliptic motion was anything but easy. I doubt whether it could subjectively have seemed less difficult than discovering the rules of structure in modern particle physics. In both these cases, separated in our science by many centuries, the procedure is the same—first find the empirical facts, then find systematic regularities, then find the reason for the regularities.

A quick answer to the question of why the work of Copernicus was so important would be to say that by placing the Sun at the centre of the solar system he made it easier to arrive at a substantially correct description of planetary motion—that it proved more straightforward to grapple with the problem that way. I think this answer contains a measure of truth. Certainly it was easier a century later for Newton and his contemporaries to tackle the dynamical problem in terms of a heliocentric description. But such an answer overlooks the pre-Copernican situation in which the geocentric theory of Ptolemy had proved more successful than the heliocentric theory of Aristarchus. Until Copernicus, experience was just the other way round. Indeed Copernicus had to struggle long and hard over many years before he equalled Ptolemy, and in the end the Copernican theory did not greatly surpass that of Ptolemy. I say this not to diminish the achievement of Copernicus but in the hope that the five-hundredth anniversary of his birth will prove an opportunity for the modern world also to pay tribute to the greatness of Ptolemy.

Let us see in general terms what was achieved. To have any hope of representing the motion of a planet in an approximate but controlled way it is necessary to know the synodic period to high accuracy.* By a controlled way I mean one free from cumulative errors, errors that start by being small but which pile up inexorably to larger and larger values as time goes on. So long as we arrange our theory so that each planet completes its orbit in the sky in the correct period there will be no such progressive errors. The errors will be of detail, resulting from our inaccuracies in describing the orbit itself. To obtain substantially the correct periods of the planets was easy, at any rate to the whole human species if not to an individual astronomer. All that need be done is to observe a planet over a large number of revolutions. We judge a revolution to be completed when the planet returns to the same position with respect to the stars. Our observation in this respect need not be very accurate provided sufficiently many revolutions are observed. Suppose we can judge position with respect to the stars within 0.5°, using simple naked eye observations. This is about $\dfrac{1}{700}$ of a revolution. So if we observed just one revolution we could expect to mistake the orbital period by about one part in 700. But after n periods we shall determine the period to within one part in $700n$, which for n large becomes very accurate. Now the planets have been observed from this point of view from time immemorial and good values for their periods were

* The synodic period is the period of revolution as seen from the Earth.

available to all makers of theories throughout historic times, to the Greeks as well as to Copernicus.

With this in mind we see that a lack of knowledge of the precise orbit, and of the dynamical behaviour in the orbit, can only result in detailed errors. At any given moment a planet will be observed to be out of its theoretical position by some moderate angle, the amount

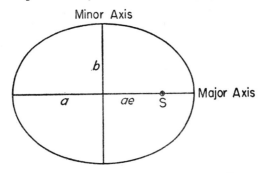

FIGURE 2. The planetary orbits are ellipses, being less flattened than the ellipse shown here. Even when the ellipse flattens only slightly, the focus S moves markedly away from the centre. It is this property which caused glaring discrepancies to appear in approximate theories, particularly when applied to the planet Mars

depending on the quality of the theory. It is interesting to classify approximate theories according to their relation to the eccentricity of the actual elliptic orbit. In Figure 2 we have an ellipse representing a planetary orbit, with semi-major axis a, semi-minor axis b, and eccentricity e. The Sun is considered to be at the focus S.

It is the nature of an ellipse that the lengths of the axes are related to the eccentricity by

$$\frac{b^2}{a^2} = 1 - e^2. \tag{1}$$

In the particular case $e = 0$ the axes are equal, $a = b$, and we have a circle. As e increases, approaching to unity, the ellipse becomes flatter and flatter. No cases with e approaching unity are found among the planets, although such cases occur often for comets. The situation for the six planets known to Copernicus and to the ancient world is shown in the following table:

TABLE 1

Planet	Eccentricity	Major axis *(in terms of Earth)*
Mercury	0.2056	0.387
Venus	0.0068	0.723
Earth	0.0167	1.000
Mars	0.0933	1.524
Jupiter	0.0484	5.203
Saturn	0.0558	9.539

In all cases the orbits are nearly circles. Even for Mercury the shorter axis is only about 2 per cent less than the larger axis. This was a most fortunate circumstance since it permitted approximate theories to be developed that were based on circles.

If one makes the approximation of replacing the elliptic orbit by a single circle with centre at the Sun the angular errors between the observed positions of a planet and the theoretical positions will be of the order of the eccentricity. This is in terms of the unit of angle, which is a fraction $\frac{1}{2\pi}$ of a full revolution, about $57°$. For Mercury, the planet in our table of largest eccentricity, the errors as seen from the Sun turn out to average about $15°$. As seen from the Earth, the errors will be less than this, however, because Mercury is nearer to the Sun than it is to the Earth. Seen from the Earth the errors will average more like $5°$. For Mars, the planet in our table with the next largest eccentricity, the errors average about $7°$ seen from the Sun. But at opposition, Mars approaches much closer to the Earth than the Martian distance from the Sun, and this greatly exaggerates the error for a terrestrial observer, to well over $10°$. Thus for an observer on the Earth errors will be largest for Mars. It is curious that although the actual orbits do not differ in shape much from circles the errors of a circular model can nevertheless be quite large.

Indeed errors as large as this were quite unacceptable to Greek astronomers of the standard of Hipparchus and Ptolemy. It was this, rather than prejudice, which caused them to reject the simple heliocentric theory of Aristarchus. Little is known of the circumstances in which Aristarchus put forward his theory. There are indications that the approach may have been somewhat tentative, which may well have been so in view of his likely knowledge of the resulting discrepancies between the theory and observation.

The planet Mars has played a great role in the history of astronomy. Quite apart from its importance in the Aristarchus–Ptolemy–Copernicus–Kepler story, it served to cast serious doubt on an earlier theory, the homocentric spheres of Eudoxus. It is worth bringing this point to mind because it has a relationship with the humanistic intellectual activity that was sweeping through Europe at the time of Copernicus. All the spheres of Eudoxus had centre at the Earth. The outermost sphere had the stars attached to it, being otherwise dark. This outermost sphere rotated to give the diurnal motion of the heavens. Inside this sphere were many other translucent ones. The Moon and the Sun were given nests of three each, while the planets each had nests of four spheres. The outer sphere of a nest had the same motion as the sphere of the stars. The second outermost of each nest was attached to the outermost by a polar axis about which it was free to turn. The third one was attached to the second in a similar way, and so on. Finally, the planet, or the Sun or Moon, was attached to the innermost sphere of the nest. In this way complex motions of the various bodies could be produced. The mathematical problem was to choose the polar axes and their points of attachment and the motions of the spheres in such a way as to reproduce the observed behaviour of the Sun, Moon, and planets. It will be clear that this theory contained so many parameters that a fair measure of success for it could be expected. But it failed in a critical respect. Because the spheres were all centred at the Earth, the motions had perforce to be of such a kind that any particular planet, or the Sun or Moon, remained always at the same distance from the Earth. Hence each planet should maintain a constant brightness—which is not true. Particularly for Mars there are large variations of brightness, because the distance of Mars changes markedly, being some five times greater at solar conjunction than at opposition.

Even so, the theory of Eudoxus found great favour with Aristotle, who extended it himself until it came to involve more than fifty geocentric spheres. This was the theory that came to the notice of the European humanists during the fifteenth century. Access to Greek culture, taken with the respect accorded to the teachings of Aristotle, contrived to draw an enormous red herring across the path of the serious astronomer. Copernicus, in particular, had to fight a war on two fronts. On the one hand he had to do fierce battle with the theory of Ptolemy—which was of an altogether higher level of sophistication than the theory of Aristotle—and at the same time he had to combat the prejudice towards a geocentric theory which was being so strongly supported by the popular adherence to Aristotle. The theory of Ptolemy, a few minor imperfections apart, worked correctly to the first order in the planetary eccentricities. Copernicus with his heliocentric theory had to do at least as well as this, which meant that *he had to do much better than the simple heliocentric picture of Aristarchus.*

Successful in this, the outcome was the heliocentric theory published in *de revolutionibus*.

In Chapter IV we shall find that, judged in terms of predictive quality, there was little to choose between the new heliocentric theory and the theory of Ptolemy. The new theory was crucially important, however, just because of the intense emotional reaction to it. Emotional opposition was so strong that astronomers and astronomy came very much into the popular eye, whether favourably or adversely is not relevant. The essentials were that astronomers became impelled to observe the planets with increasing accuracy, and funds to enable them to do so were forthcoming.

The key issue lay, once again, in the motion of Mars. By working correctly to the first order in the eccentricity, the discrepancies between theory and observation were reduced to terms of the order of the square of the eccentricity. But this gave a maximum error for Mars of more than $1°$, which was still very glaring. Persistent discrepancies for Mars, taken with the highly charged controversial situation, forced an attempt at a more far-reaching theory. The time became ripe for increased understanding.

The Copernican theory fitted the planets of small eccentricity, Venus, Earth, Jupiter, and Saturn very well, to within the errors of observation in some cases, so there had to be something 'right' about it. The theory thus provided a base from which to proceed. Progress was certain as soon as a man, or men, of adequate stature came along. As we know in retrospect, favourable circumstances came with the observations of Tycho Brahe and with the appearance of Kepler. Kepler fitted the task of advancing the theory, not only in stature, but also in the demonic energy with which he tackled the problem of Mars. The sheer volume of calculation which he carried through fits him to be called the iron man of science—much as Richard Wagner may be called the iron man of music.

In the beginning Kepler contented himself with setting right one or two minor blemishes in the Copernican theory. Finding this did little to help the problem of Mars, he then began a long series of trial calculations, which in effect were a search, along Copernican lines, for the missing second order terms in the planetary eccentricities. Kepler found improvements, but not complete success, and always at the expense of increasing complexity.

Kepler and his successors might well have gone on in this style for generations without arriving at a satisfactory final solution, for a reason we now understand clearly. There is no simple mathematical expression for the way in which the direction of a planet—its heliocentric longitude—changes with time. Even today we must express the longitude as an infinite series of terms when we use time as the free variable. What Ptolemy, Copernicus, and Kepler in his early long calculations, were trying to do was to discover by trial and error the terms of this series. Since the terms become more complicated as one goes to higher orders in the eccentricity, the task became successively harder and harder. The modern expression for the heliocentric longitude θ measured from perihelion in terms of the time t is

$$\theta = nt + 2e \sin nt + \tfrac{5}{4}e^2 \sin 2nt$$
$$- e^3[\tfrac{1}{4} \sin nt - \tfrac{13}{12} \sin 3nt] + \cdots \quad (2)$$

Here e is the eccentricity and n is the planet's 'mean motion', defined in terms of the sidereal period P by

$$n = \frac{2\pi}{P}. \quad (3)$$

The heliocentric distance r is related to time by

$$\frac{r}{a} = 1 - e \cos nt + \tfrac{1}{2}e^2(1 - \cos 2nt)$$
$$+ \tfrac{3}{8}e^3(\cos nt - \cos 3nt) + \cdots \quad (4)$$

The forms of the terms in these series are not easy to guess, and if guessed would not seem at all natural or

TABLE 2

Planet	Copernican value of a	Modern value
Mercury	0.3763	0.3871
Venus	0.7193	0.7233
Earth	1.0000 (standard)	1.0000
Mars	1.5198	1.5237
Jupiter	5.2192	5.2028
Saturn	9.1743	9.5388

elegant. Notice that one has to guess *both* series in order to be able to relate predicted heliocentric values to the observed geocentric longitudes of a planet. As well as knowing P one must also know a and e, although the semi-major axis a need not be known in the usual local units (metres). It is sufficient to know the values of a in relation to one another, for example in relation to a for the Earth. Copernicus had tolerable values of P, reasonably good values of e, and quite good values of a.

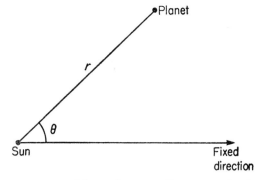

FIGURE 3. The polar coordinates r, θ of a planet taken with respect to the Sun. The fixed direction here is to be interpreted as fixed with respect to distant parts of the universe (see *Epilogue*)

To come back to Kepler, the stroke which permitted Kepler to solve the problem came from abandoning a frontal attack on the problem of guessing the two series given above. Instead Kepler set himself the problem of determining the *shape* of the orbit, which is to say determining r, not in terms of time, but in terms of θ (cf. Figure 3). With this approach the answer was elegant and reasonably straightforward, r varied with θ like

$$\frac{1}{1 + e \cos \theta}. \tag{5}$$

Kepler knew immediately that this relationship defined an ellipse and that the constant of proportionality between r and (5) had to be related to a and e in the usual mathematical way, namely

$$r = \frac{a(1 - e^2)}{1 + e \cos \theta}. \tag{6}$$

There remained the problem of finding how a planet moved in a *now known* orbit. This problem was solved by another remarkably elegant discovery, a planet sweeps equal areas of its ellipse in equal times, the time for the whole ellipse being just the known period P.

Another remarkable discovery was that the empirically determined values of P and a were such that P^2 was proportional to a^3, a fact which could have been known to Copernicus but which does not seem to have been noted by him. This relationship remained empirical, as did Kepler's other discoveries, until a dynamical theory became available eighty years later.

So quite suddenly the apparently intractable problem of describing the planetary motions was made to look simple. From the moment Kepler found equation (6) the course of modern science was set. Until then there was nothing really firm for science to get its teeth into. Wipe out all that has been discovered since Kepler. Go back to the year 1610 and do it all again. Let there be no Newton this time, no Euler, no Lagrange, ... Yet there will be others making the same discoveries, perhaps more slowly but probably not much more slowly. The flowers along the way may be different but the path will be the same. It is because Copernicus focused the attention of the world at precisely the right spot—the place where Nature simply had to give up her secrets—that today we judge his work to have been so important. In mountaineering jargon, Copernicus found the 'point of attack'. Let us now go back to his own times, and see how he lived and worked, and what manner of man he seems to have been.

II
Mr Standfast

THE NAME AS WE KNOW IT, NICOLAUS COPERNICUS, is a latinized self-name, adapted for scholarly writings from the family name of Niklas Koppernigk. Copernicus, as I shall continue to call him, signed himself Coppernic on official documents, closer to the Koppernigk form. The family seems to have migrated from Silesia, moving into Poland early in the fifteenth century. It was commonplace for settlers from the west to move into Poland throughout the fourteenth and fifteenth centuries. Indeed, the Teutonic Knights, an Order founded during the Crusades, were invited into Poland already in the thirteenth century, to help it is said with certain territorial ambitions of the Polish rulers. For their services the Knights were given the eastern Baltic region which we know today as East Prussia. Between East Prussia and Poland lay the little province of Warmia, destined to play a critical role in the life of Copernicus.

With its northern flank reasonably secure, Poland in the latter part of the fifteenth century was pressing outward over a wide arc, from Czechoslovakia and Hungary in the south-west, to Russia in the east. Poland controlled a long wide corridor stretching south from the Baltic almost to the Black Sea.

Such a broad pattern of life could hardly have failed to have had influence on Copernicus. To grow up in an era of free movement of peoples is an advantage towards acquiring an unprejudiced mind, always assuming the times are not so violently disturbed as to make sheer physical survival a principal factor. Copernicus was doubly lucky here. He acquired an independence of outlook while enjoying the security of a stable family. One Johann Koppernigk, appearing in the records of Cracow as merchant and banker, was probably the father of Niklas Koppernigk, also merchant and banker. Niklas moved from Cracow to Torun (Plate 2) on the Vistula, and it was here that the second Niklas Koppernigk, the astronomer, was born on the 19 February, 1473. With both father and grandfather immersed in the field of commerce and banking, it should not surprise us to find Copernicus in later life writing a well-reasoned monograph on practical economics, which in the outcome had a significant effect on the money systems of Poland and East Prussia.

Niklas the elder died when the future astronomer was only ten. Thus it came about that Copernicus was adopted by his uncle, Lucas Watzenrode (Plate 3), who six years later was to be appointed Prince-Bishop of Warmia. Since Watzenrode appears to have been a man of high temper it is something of a mystery how he came to be appointed to such a post. At all events he came to a position where he could be of great material help to Copernicus, who was never to know the crushing disability of economic privation.

Copernicus was soon travelling to Cracow, on his uncle's advice or instruction, to begin studies at the university there. He was fortunate in that Adalbert Brudzewski (1445–97), professor of astronomy at the university (although moving from astronomy to philosophy at about this time), had been a pupil of Johann Müller (Regiomontanus, 1536–76) who was himself the follower of Georg Puerbach (1423–61) at the University of Vienna. Thus Puerbach's book *Epitome in Ptolemaei Almagestum* probably became known to Copernicus at the very beginning of his career. During these early student days the university is said to have received a present of astronomical instruments, primitive indeed by our standards, but very up-to-date at the time. Probably Copernicus played a part in setting up these instruments and in making observations with them, for a few years later when taking up studies in Italy we find him working with Maria da Novara (1454–1504), professor of astronomy at Bologna, 'not so much as a pupil as an assistant'. This paved the way for the observational work which Copernicus was to need in his later years.

A truly momentous event occurred during these Cracow days. Copernicus was nineteen in the year Columbus discovered America. We have no account of the impact this discovery made on young Copernicus, but it needs little imagination to understand how it came about that he turned so wholeheartedly to the views of the Greek, Heraclides, who first suggested that the Earth is a sphere and that the apparent diurnal rotation of the heavens is due to rotation of the terrestrial sphere, not to an actual motion of the heavens themselves. It is indicative of how far medieval Europe had advanced beyond the Greeks in the understanding of practical physical problems that Copernicus, when confronted by the old argument that rotation would cause the Earth to fly apart, answered by remarking how much more surely would the heavens fly apart if called on to rotate at a similar rate. Here we have the sound physical perception that the forces developed for a given rate of rotation are greater in a large system than in a small one. Although *de revolutionibus* is concerned mainly with abstract geometrical constructions, it is as well to remember that behind all this formal work lay a great deal of physical common sense. Indeed common sense, or sound sense as we might more properly say, was the epitome of Copernicus' life.

Again on his uncle's advice we find Copernicus travelling in 1496 from Poland to the University of Bologna. Together with his brother, Andreas, he is said to have walked across the Alps. It is a matter for regret that neither brother apparently saw fit to commit an account of the journey to paper—mountaineers as well as astronomers would richly prize such a journal.

In 1497, through Lucas Watzenrode's good offices, Copernicus was appointed a canon of the Chapter of Warmia. Lucas would seem to have found this nepotism not altogether easy, for he brought it off only at his second attempt. Copernicus now had a measure of financial independence which he was to retain for the rest of his life. Andreas had to wait two more years before receiving his canonry, but Andreas was not to profit from the appointment in the same degree as Niklas, for he was to die young—possibly from leprosy.

Copernicus returned to Warmia in 1501, to be inducted into his canonry as it is usually said. Then he was back in Italy, now at Padua studying medicine. This is usually thought to have been at his uncle's instigation. However, Copernicus' mother, Barbara, was ill and indeed nearing the end of her life at this time, which would seem to give a more plausible reason for the return to Poland—and more than one young man has been impelled into the study of medicine through the illness of his mother. Copernicus returned finally to Poland in 1503. His student days were truly over at last. Thirty years old now, he had studied mathematics and astronomy, medicine, and law, he had a knowledge of many languages, latterly including Greek, and he was a master of most of them. This was the pattern of the Renaissance man. It is a pattern which many educators (the devil stir them) seek after in our own time. You were fortunate if you were born a man of many parts in the Renaissance, but the child born today with abilities in many fields is far less fortunate than it may seem at the onset. He will be encouraged by the educators to develop 'broadly'. Likely enough he will please his teachers by turning in good performances over a wide range of subjects, but he will end up with the sad discovery that there is no field in which he can compete successfully with the narrow specialists whom our educators deplore so loudly. The chances are that the widely gifted child will wind up today as an embittered person without real self-confidence. But it was otherwise during the Renaissance, and it was very much otherwise with Copernicus. He could afford the many years of student life because there were no narrow specialists treading on his heels. Copernicus even made a stab at the arts, but Plate 4, based on a self-portrait, shows him to have been no Leonardo da Vinci.

How far towards the *de revolutionibus* was Copernicus at the end of his student days? We do not know and can only speculate. At some time before *de revolutionibus* Copernicus sent out a manuscript *Commentariolus* in which he describes in a kind of extended abstract the ideas of his heliocentric theory, leaving aside for a later larger work the more difficult mathematical demonstrations. Although certain of the constructional methods used in *Commentariolus* are different from *de revolutionibus*, it is clear that *Commentariolus* was written after Copernicus had arrived at the main structure of his theory. So if we knew the date of *Commentariolus* we should be well along the road towards answering our question, but one scholar has given a date post-1530 and another pre-1512, which covers quite a range of possibilities! If we accept the earlier date much of the work must have been done before the return to Poland in 1503, since on his return Copernicus became secretary, legal adviser, and personal physician to his uncle. He resided until 1512 at Heilsberg (Plate 5) where he became involved in many a sharp political situation. Lucas Watzenrode, a tornado of a man, was inclined to deal hot-headedly with political problems. We find Copernicus calming the old fellow, we find him writing reports and memoranda, advising on negotiations, acting as physician—and in earnest too, since his uncle was not a fit man. We find him travelling, locally and nationally. These activities are not of a kind to go well with the pondering of a deep intellectual problem. I believe then, that if Copernicus, by 1509 or thereabouts, was in possession of the guts of the solution to the heliocentric problem, of necessity he must have had most of it already in 1503. In which case I find it surprising that he left no trail of curiosity and wonderment behind him in Italy.

In 1509 Copernicus published the first direct translation from Greek into Polish, a collection of ribald letters of the Byzantine poet Theophilactus Simocatta, the book being dedicated to Lucas Watzenrode. This translation does not fit with a man concentrating deeply on an abstract mathematical problem. If Copernicus was really immersed in his theory over the hectic period

PLATE 1. The motions of the planets over a period of some twenty years have been simulated to show how they appear to an observer on the Earth. (*Münich Planetarium*)

PLATE 2. Copernicus was born in Torun on 19 February 1473

PLATE 3. Lucas Watzenrode (Watzelrode), uncle of Copernicus, appointed Prince-Bishop of Warmia (*ca* 1489). The election
was a snap affair, hurried through in less than a week. The
Chapter (Diocese) of Warmia was anxious to secure autonomy
in the appointment of its Bishops. The choice of Watzenrode
was expected to be acceptable to the Polish King. While this
may have been true in a personal sense, the manner of his
election was to be a source of contention throughout the rest
of his life. The squabbles over this issue involved Copernicus
closely in the years from 1503 to 1512

1503–12, surely he would have snatched every spare moment for his astronomical investigations. He would hardly have had time to spare for Theophilactus. It is a curious thought that but for the almost accidental circumstances which seem to have led to publication of *de revolutionibus*, Copernicus would have gone down as the author of two books, one of ribald letters, the other a dissertation on money policy.

We should be grateful for the matter of these ribald letters, because this book of 1509, and an episode from Copernicus' later life, provide us with rare glimpses of the man in flesh and blood. Except in this one moment I cannot hear Copernicus laugh. I hear Kepler chuckling away throughout his life. I hear great gusts of laughter from Galileo. I get none from Newton, though, only sniffs, because I do not feel there was any laughter in Newton as a person. For Copernicus the situation was different. The portraits of the fifteenth century stare down at us from the walls in impassive masks. The fifteenth century was an age in which men learned to keep their true selves hidden from the world, especially men with the responsibilities now falling on to the shoulders of Copernicus.

To give my own appraisal of the astronomical situation, I suspect that Copernicus had arrived at his position concerning the rotation of the Earth in his university days. I suspect Copernicus, as a student, became acquainted with a publication *An terra moveatur an quiescat* in which Johann Müller (Regiomontanus) sets out with approval all the old arguments against the rotation of the Earth, in which he comments sarcastically on the idea of the Earth being roasted like meat on a rotating spit. If Copernicus had decided already in favour of the Earth's rotation, say after Columbus in 1492, he would have felt cause to doubt all the other arguments of Regiomontanus. In the manner of the young, he would have been inclined to regard the old man as a bit of a fool and he would have come to discount Müller's arguments in favour of the Ptolemaic theory. I see the young Copernicus with the conviction that one at least of the main ideas of the old theory was wrong and if one of its cornerstones was gone perhaps the old theory was wrong as a whole?

After his uncle's death in 1512, Copernicus at last took up his canonry at Frauenburg, a town at the mouth of the Vistula (Plate 6). The Cathedral there overlooks a freshwater lagoon opening into the Gulf of Danzig. This was a drastic change, in all material and outward respects. Instead of his being in many ways the most important man in Warmia (the secretary of any organization is usually its most important man, *de facto* if not in name). Copernicus was now just a canon among many canons (there were fifteen others) domiciled in a not particularly important town. There can be little doubt that such a change must have involved a considerable drop in what is nowadays called the 'pecking order'. Biographers have made much of the emergence of Copernicus in this period as a physician of widespread renown, eagerly sought after by both high born and low. Such an attempt to represent his life as the unimpeded voyage of overriding genius seems questionable. Unless the men of his day were very different in emotional make-up from those of our own, officials at the Headquarters of the Chapter of Warmia at Allenstein must have grown restive—to say the least—at the influence which Copernicus exercised over his uncle, Lucas Watzenrode. Officials never relish a situation in which a youngish man from outside comes to be promoted above their heads, as Copernicus was in the years from 1503 to 1512. In such circumstances it is not the way of officials to move openly, they lie low biding their time. When their time comes, as it did with Watzenrode's death, they step in to take revenge. It is in such a light that I interpret the removal of Copernicus to Frauenburg. And I interpret the medical activities of Copernicus as a reaction to demotion. By the respect accorded to him as a medical authority he contrived to maintain his reputation as a man of consequence.

To take this attempted interpretation a bit further, I would see something of an ambivalent attitude developing at Allenstein in respect of Copernicus. Immediately following Watzenrode's death I see the situation as I have just described it. But after a few years had slipped away, after immediate hostilities had weakened, it would be remembered how very able Copernicus was. The authorities at Allenstein, like cats in a box, would have found new quarrels. There would perhaps be some among them who would find themselves contemplating the possibility of his return, not to a top level job, but to something a bit hum-drum requiring precise competent administration. It is in these terms that I see the recall of Copernicus in 1516 to the job of administering certain outlying estates of the Chapter. This was a far cry from the big issues with which he had been concerned during his uncle's lifetime. It is of interest that he accepted such a post at all.

The storm clouds of war were gathering during the years Copernicus occupied this minor position. In 1520 the Teutonic Knights under Albrecht von Hohenzollern (1490–1568) attacked Poland across Warmia in an attempt to secure independence for East Prussia. One of few positive aspects of such a situation is that men of ability rise rapidly. Petty animosities become forgotten in the urgency of the moment. So it was with Copernicus. We find him administering outlying estates no longer. By 1519 he was back essentially at his old job, as Secretary and Chancellor of the whole Chapter of Warmia. I see no reason to doubt the story that, when the Knights besieged the capital of Allenstein, senior officials fled the city, leaving Copernicus behind as Governor-in-Charge. It is consistent with all we know of his ability and character that the city did not fall.

Nor were the Knights successful in the war generally. The outcome was devastation and inflation throughout Warmia. Copernicus was called on to prepare a memorandum on the state of the country, a document which eventually served the Polish King as a basis for peace negotiations. It was at this time that Copernicus became deeply concerned with problems of economics. He became involved with farmers, helping them rebuild and restock their herds. He sought to maintain fixed prices and to maintain standards for staple commodities. In 1523, in an interregnum between Bishops, he acted as Administrator-General of all church property in Warmia—a far cry indeed from the minor post of 1516.

The merchant banking skills of the Koppernigk family came to the fore in this period. Copernicus wrote a tract, *Monete cudende ratio* in which he proposed currency reforms, some of immediate good sense, others far-reaching into our modern world. Up to this time, different cities in Poland and Prussia had each minted their own coinage. This produced a situation analogous to that which occurred in modern Germany after the last war. [The Allies agreed at Yalta to use a common currency. While this was sensible enough, neither Roosevelt nor Churchill insisted on defining how much currency each power should print. The inevitable result of this omission was that the Russians swamped the market, inflating money to whatever degree was necessary to prevent German recovery.] To permit normal trade and interchange between cities there had of necessity to be a mutual recognition of coinage values. Yet this permitted cities producing large quantities of debased coins to override cities which sought to maintain standard quantities of silver and gold in their coins. This indeed was the origin of Gresham's law 'that the bad drives out the good', a law which Copernicus recognized some thirty years ahead of Gresham. The clear-cut solution which Copernicus proposed, and which eventually came to be accepted both in Poland and East Prussia, was to replace this individual minting of coins by properly controlled national currencies. The tract contained other straightforward proposals for reform, but it also contained a proposal to control the economy by regulating the supply of money—an idea which carries us into the very forefront of economic controversy in our modern world.

The war with Prussia made Copernicus an important person in his own right. He could look now for high office if he wanted it. Yet following his term as Administrator-General we find him returning to Frauenburg. To understand why he should have preferred the comparative obscurity of this little Baltic town we must think again of astronomy. Let us go back to the years of 'exile', 1512–16. Copernicus was faced in these years, not just by the material 'downbeat' we have discussed above, but also by an emotional downbeat. Copernicus lost his father at the age of ten. The father–son relationship quite evidently became transferred to Lucas Watzenrode, but now Watzenrode was also dead. So too was Copernicus' mother. By this time, Andreas, his companion in the walk across the Alps, was sick with an incurable ailment. Of his sisters, Barbara had retreated from the world into a nunnery. Only Katherina, married to a Cracow merchant, effectively remained of his family. Copernicus in 1513 must have felt very much alone. Only ten years earlier he had returned from carefree student days in Italy. Now, after a decade of hectic activity, the world had suddenly taken on a darker hue.

It appears likely to me that the year 1513 marks a decisive turning point in the life of Copernicus. It is to the period 1513–16 that I would place the first work in astronomy *at a fully serious level*. I suspect this may well have been the first time in his life when Copernicus set himself to meet the full quantitative challenge of the Ptolemaic theory. To this point, I would expect him to have developed a clear qualitative insight into the planetary problem but I think the guts of the Copernican theory as we know it still remained to be worked out. I see the situation much as I see the deafness of Beethoven. I do not believe Beethoven would have developed the full measure of his greatness had he not become deaf. I similarly doubt if Copernicus would have driven through the more difficult aspects of his theory had it not been for the adversities which befell him in the year 1513.

It is worth considering objections to this point of view. Why in this case did Copernicus accept the minor administrative post in 1516? There are several good answers to this question. Firstly, as a matter of self-respect. Copernicus must have wanted to prove himself in the world of affairs, to prove that he could succeed without the patronage of his uncle. Secondly, the post being a minor one, would not interfere too seriously with astronomical work. Thirdly, the storm clouds of war were gathering. Copernicus may have wanted to 'do his bit', as many scientists did in the last war. Certainly the increasingly heavy involvement over the years 1519–23 can be explained in these terms. It is not unreasonable to suppose that the war prevented Copernicus from following his intentions until after 1523. Then, one can fairly argue, he felt able to return to his personal researches—and this was why he elected to return to Frauenburg, instead of continuing at Allenstein. It is to the following decade, 1523–33, that I would attribute the quantitative core of the Copernican theory.

In 1513, Copernicus was already forty. There may be critics, mindful of the modern adage which says that the world's best theoretical physicists are all worked out at thirty, who will consider forty to have been too great an age for Copernicus to have begun an assault on the most crucial aspects of his theory. To this criticism I would reply that if modern physicists are worked out at thirty it is because they have grossly overworked in earlier years, and that the condition of being 'worked out' consists of boredom and lack of energy, not of lack of ability. Newton (1643–1727) wrote his *Principia* when he was about forty. Schrödinger was approaching forty when he found his wave equation.

Beethoven was in his fifties when he wrote what many feel to be his finest music and Verdi was twice the age of forty when he wrote the tremendous trumpet passage at the end of *Otello*.

Let us proceed then on the basis that Copernicus arrived at the qualitative aspects of his theory in the period up to 1503, that he began serious quantitative work only in 1513, and that the full measure of the theory finally took shape in the decade from 1523. This places 1533 as a likely approximation to the date of *Commentariolus*, a dating which fits the known facts in a reasonable way. It may well have been for Johann Widmanstadt, secretary to Pope Clement VII, that *Commentariolus* was written. Widmanstadt is known to have presented Copernicus' ideas to Pope Clement in 1533. In 1536, as an outcome of discussions within the Church, Copernicus received a letter from Cardinal Schönberg urging that a full account of the theory be published (Galileo rest in peace). Word of the new theory also reached Protestant circles. In the Spring of 1539 Copernicus received a visit from Georg Joachim von Lauchen (Rheticus). In Wittenberg, von Lauchen had heard a synopsis of the theory. His interest being aroused, he decided to learn more of it from Copernicus himself. Intending at first to spend but a few weeks or months in Frauenburg, von Lauchen in fact spent two years there. It is usually said that von Lauchen occupied himself over this extended visit in familiarizing himself with the detailed manuscript which was eventually to be published as *de revolutionibus*.

In the light of these facts it is hard to accept an early date for *Commentariolus*. It scarcely seems possible for Copernicus to have sent out a quite substantial account of his theory as early as 1512 and for it to have produced no significant reaction until 1533.

The circumstances in which *de revolutionibus* came to be published have often been discussed. It is generally accepted that Copernicus was exceedingly reluctant to publish. Von Lauchen is usually credited with persuading him to do so. Yet high dignitaries of the Church were suggesting publication long before von Lauchen appeared in Frauenburg—the letter from Cardinal Schönberg was dated November, 1536—and Tiedemann Giese, a close and valued friend, was also firmly urging publication. It would seem then that the factors which deterred Copernicus must either have been scientific or were personal to himself. Possibly von Lauchen played a part, which neither Schönberg nor Giese could play, of allaying certain scientific misgivings. I will return to this interesting possibility after considering the more personal issues.

Copernicus wrote in July 1540 to Andreas Osiander asking if it would be possible for the theory to be published without exciting hostile criticism. Osiander replied in the spring of the following year: 'For my part, I have always felt about hypotheses that they are not articles of faith, but bases of calculation, so that, even if false, it does not matter so long as they exactly represent the phenomena of the celestial motions . . .' From this interchange one detects a fear of criticism. On another occasion Copernicus remarked quite explicitly that he was 'afraid of scorn and contempt on account of the novelty and inconceivable nature of his theory'. But scorn and contempt from whom?

A new Prince-Bishop of Warmia, Johann Flaxbinder (John Danticus), was elected in 1537. It seems likely that Copernicus played a considerable role in this election, opposing the new Bishop in favour of another candidate—possibly Tiedemann Giese. Copernicus then became subject to petty persecution and to a particular prohibition which appears to have affected him deeply. Anna Schilling, daughter of a cousin of Copernicus' mother, a widow and a woman of considerable culture, had come to Frauenburg to act as his housekeeper. On moral grounds, the new Bishop insisted she must leave. Copernicus fought the case to the limits of what was possible, but his petition to retain Anna was not successful. We see vividly in this issue the extent to which Copernicus was not his own master in matters we today would take for granted. Copernicus was immersed in a local world, not a world with modern ideas of freedom, not even the larger world of Rome, and in this small local world he was unwilling to be made to seem a fool. It is possible that his reluctance to publish arose at no deeper level than this.

By this time Copernicus was in his middle sixties.

There was no medical service in those days capable of setting to rights all the small defects of the body which necessarily accumulate in all of us at such time of life. The possibility is that Copernicus by 1538 was just plain tired and that the final revision of his manuscript represented a formidable and daunting task. Nevertheless, he agreed at last to undertake it. It is not clear whether this decision was taken before or after still another abstracted version, *Narratio prima de libris revolutionum*, was sent by von Lauchen in 1540 to his old teacher Johann Schöner in Nuremberg. The *Narratio prima*, written presumably with Copernicus' consent, appears to have had a favourable and somewhat sensational reception. Possibly this could have influenced Copernicus finally towards publication.

Von Lauchen extended his residence in Frauenburg into 1541, in the hope that eventually he would be able to return to Wittenberg with the finished manuscript. But it is usually said that he left Warmia with only early mathematical chapters. The story now gains a peculiar momentum. Copernicus is said to have entrusted his manuscript to Tiedemann Giese, to make what arrangements he considered best. Giese is said to have sent it to von Lauchen. Then we are told that von Lauchen, having left Wittenberg to take up a post in Leipzig, entrusted Andreas Osiander to see the work through the press. With the faithful von Lauchen thus removed, Osiander surreptitiously removed the Preface written by Copernicus himself and substituted one of his own making, the Osiander Preface being in substance the same as the Osiander letter to Copernicus in 1541, '. . . hypotheses are not articles of faith . . .'

As the late Walter Baade used to say, one can detect 'a slight odour of rat' in this story. Are we to believe that von Lauchen, after making the long journey from Wittenberg to Frauenburg in the hope of learning something of the Copernican theory, after waiting there two years in the hope of acquiring the precious manuscript, would suddenly turn it over to Osiander? Is the journey from Leipzig to Nuremberg so great that von Lauchen could not have made that particular journey, even on a number of occasions? And is it not curious to find Copernicus writing to this same Osiander (1540) before there was any question of Osiander's involvement with the manuscript?

The curiosities do not end here. By a lucky stroke Copernicus' own manuscript, after being thought lost, was discovered in Prague in the second half of the nineteenth century. It became immediately clear that *de revolutionibus* was not set from this manuscript. Quite a major editing job had been done between the manuscript and the printed version, editing which in retrospect must be considered improper. The manuscript is untitled so we do not really know how Copernicus wished his book to be named. From additions to the text it has been speculated that the technical editing was done by a certain Erasmus Reinhold. But what has happened in the meantime to von Lauchen?

The technical editing could have been done just as well in Leipzig as in Nuremberg.

In a first departure from this unsatisfactory story, I tried the idea that von Lauchen stayed for two years in Frauenburg to help Copernicus prepare a final manuscript. Suppose his help proved less than competent and that in the end Copernicus had to seek assistance elsewhere—Osiander and Reinhold. This would explain many of the apparent contradictions. But an examination of the *Narratio prima* shows more brashness than incompetence. I also found the amendations and corrections in the Copernicus autograph to be firm and decisive, which seemed inconsistent with the idea of his having become too enfeebled to work alone.

At this stage I became impressed by the general similarity between the *Narratio prima* and *Commentariolus* (alternative geometrical constructions are not important here). If a brash young man had not hesitated to make what was essentially a copy of *Commentariolus* would he not also have made a copy of the main work? On this basis, von Lauchen would have left Frauenburg after his two year stay, not just with the early mathematical chapters of *de revolutionibus* but with his own personal copy of the whole book. And now, just as he had publicized his version of *Commentariolus*, so he arranged to print his version of *de revolutionibus*.

Copernicus finally released his autograph to Tiedemann Giese, and Giese very likely passed it to von Lauchen, just as the usual story has it. But by this time the book was already being printed from von Lauchen's amended copy. On this interpretation we can easily understand the angry letter sent after the death of Copernicus by Giese to von Lauchen, complaining in strong language at the liberties taken with the book.

It is worth noticing that von Lauchen's copy would quite likely not have contained a preface, since authors usually add the preface as a last touch. It is therefore possible that Osiander did *not* suppress the Copernicus Preface. In the absence of an author's preface he added an editorial preface in which he referred to Copernicus in the third person. Somewhat naturally, his preface was similar to his letter of 1541 to Copernicus. The sentiments it expresses agree remarkably well with the outlook of modern theoretical physics, and are not at all inept, as earlier generations have supposed.

The rest of the story is quickly told. The book appeared in March 1543, but a copy did not reach Copernicus until May. He was then on his deathbed, and his actual death occurred within hours of the book's arrival.

To the citizens of Frauenburg it must only have seemed as if an infirm old man of some past distinction had passed from their midst. They were not to know that drums were already rolling with the noise of distant thunder. His fellow-canons, as they laid him in his grave, were not to know that Copernicus had detonated an overwhelming explosion of human knowledge, an explosion still with us in our own day, and whose eventual outcome we cannot yet foresee.

III
The Heliocentric Theory of Aristarchus

WE SAW IN CHAPTER I THAT THE SIMPLE HELIO-
centric theory of Aristarchus leads to unacceptably
large errors, particularly for Mars, even if we only
compare the predictions of the theory with naked eye
observations. Yet to arrive at the picture of Aristarchus
a considerable degree of sophistication is already needed.
In this chapter we shall see more closely what is involved.

There was little chance of ancient astronomers arriving
at a tolerable picture of the planetary motions until
after it had been realized that the Earth is essentially
spherical in shape. Two observations which led to this
conclusion are worth mentioning. When one journeys
south, new stars not previously visible appear above the
southern horizon, and stars dip lower towards the
northern horizon. The Greeks knew long before
Hipparchus that the constellation of the Plough comes
down to the northern horizon when seen from Egypt
but not when seen from Greece. Assuming the stars to

be distant objects, this shows that the Earth is not flat. More directly, when the Earth comes between the Sun and the Moon the shadow which the Earth throws on to the Moon is seen to be circular in shape. By judging the apparent radius of the shadow it is possible to determine the ratio of the distance of the Moon to the radius of the Earth, and Hipparchus succeeded in obtaining a surprisingly good estimate this way (the ratio is about 60). The spherical shape of the Earth was understood and accepted by the Greeks from the time of Plato, so there was no impediment to later astronomers on this score.

The diurnal rotation of the heavens is the most obvious motion to be fitted into our picture. It makes little difference from a purely kinematical point of view whether we consider the Earth to be fixed with the heavens rotating or *vice versa*, although dynamically it would be awkward (but not impossible) to think in terms of a rotation of the heavens. Indeed kinematically it is somewhat simpler, at any rate to begin with, to think in terms of the heavens as rotating. We have the situation shown in Figure 4, where we think of the stars being projected on to a distant sphere with centre taken at the Earth, the celestial sphere. The celestial sphere is considered to rotate about an axis directed towards the pole star P. The circle in which the plane through the centre of the Earth taken perpendicular to the polar axis PP_1 (Figure 4) cuts the celestial sphere is called the celestial equator. The angle which a line

drawn from our particular position on the surface of the
Earth to the centre makes with this plane is called our
latitude.

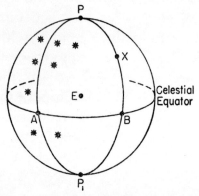

FIGURE 4. The determination of latitude and
longitude on the celestial sphere. The latitude
of the point X is the angle which the circular
arc XB subtends at the Earth E, the longitude
is the angle which AB subtends at E, A being
a fixed reference point (see Figure 6). The
celestial sphere rotates once in a day, the
reference point A partaking in the motion. It
follows that this diurnal rotation does not
change latitudes and longitudes

We can obviously set up a system of latitude and
longitude on the celestial sphere as well as on the surface
of the Earth. This is what astronomers do in order to
describe the positions of stars and other objects. Longitudes are reckoned with respect to a standard position,
just as longitudes on the Earth are reckoned with respect

to a standard place (Greenwich). Notice that the rotation of the heavens does not affect latitudes and longitudes on the celestial sphere because the standard position is taken to partake in the diurnal motion.

For the Sun, Moon, and planets we determine latitudes and longitudes in the same way, but in these cases the latitude and longitude values change with time. At any particular moment, if a planet, or the Sun or Moon, happens to be very close to a particular star we assign essentially the same latitude and longitude to it that we do for the star, but because the planet or the Moon or Sun, as the case may be, is found to move with respect to the stars we have to be constantly changing the assigned latitudes and longitudes. The determination of how the latitudes and longitudes change with time is just the problem before us.

Of all such cases it is easiest to observe the motion of the Moon with respect to the stars. This shows up in only a few nights of casual observation. By following the Moon for a month or two it is found to move along a path on the celestial sphere of the kind shown in Figure 5. Careful observation over many years shows, however, that the Moon does not quite retrace its path from one month to the next. Starting with a particular path in a particular month it takes 18.61 years before the Moon returns to that same path, and in this time the angle which the plane of the path makes with the plane of the celestial equator goes through all values from about 19° to 29°, and back again to the starting value. This complex behaviour was probably already known to stone-age man. It plays a critical role in the occurrences of eclipses and for this reason has been important to astronomers throughout the ages. It must have seemed just as important to explain this curious

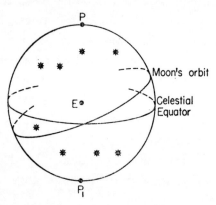

FIGURE 5. In addition to sharing the diurnal rotation of the celestial sphere, the Moon follows a path among the stars which it completes in a lunar month of 27.3 days

behaviour of the Moon as it seemed to explain the motions of the planets. Yet whereas the planetary problem lay reasonably within the grasp of early astronomers, the theory of the motion of the Moon was not accessible to them, or to Copernicus or to Kepler. The broad features of the lunar motion had to await Newton's theory, while the fine details were not subject to analysis until the nineteenth century—indeed there are problems still extant.

The early astronomers were not to know *a priori* which problems they could hope to solve. Perforce they had to make a shot at everything. And because insoluble problems were mixed up with soluble ones their task in dealing with the soluble ones was made all the harder. This must always be remembered in attempting to understand the difficulties which beset Ptolemy and Copernicus. Both expended much effort in attempting to understand the Moon whereas they would probably have gone further with less effort if they had ignored this problem.

The Sun is harder to observe than the Moon, because the glare of the Sun prevents the background of stars from being seen in full sunlight. Nevertheless, by observing stars close to the Sun immediately before sunrise and immediately after sunset, it is possible to make a determination of the path of the Sun on the celestial sphere. It is found that the latitude and longitude of the Sun also change from day to day, and that the Sun returns to its original position in a year. The path followed by the Sun does not alter significantly from year to year, so fortunately we do not have to face the complexities which bedevil the motion of the Moon. The path of the Sun projected on the celestial sphere is shown in Figure 6. The plane of the path makes an angle of about $23.5°$ with the plane of the celestial equator. It is this tilt which explains the seasons of the year. For the northern hemisphere of the Earth there is midwinter when the Sun is at B, spring at γ, summer at

A, and autumn at γ'. The points γ, γ' are the nodes of the Sun's orbit, γ being called the first point of Aries. Longitudes of stars on the celestial sphere, as well as longitudes of planets and of the Sun and Moon, are reckoned with respect to γ.

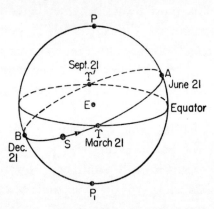

FIGURE 6. The Sun follows an annual path among the stars, the plane of the path making an angle of about 23.5° with the plane of the equator. The Sun's orbit cuts the equator at γ, γ'. The point γ is taken as the reference point for measuring longitudes on the celestial sphere. That is to say, the point γ in this figure is the point A of Figure 4

It makes no difference, kinematically, whether we think of the Sun's observed path among the stars as being due to a motion of the Earth around the Sun or of the Sun around the Earth. For the moment, then, let us follow the old belief that the Sun moves around the Earth. The path drawn in Figure 6 is a circle because it is a projection of the orbit on to the celestial sphere. The actual orbit in space need not be a circle, and indeed must not be a circle because, if it were, summer and winter would be equal, whereas the seasons are found to be unequal. It was already known to the Greeks that

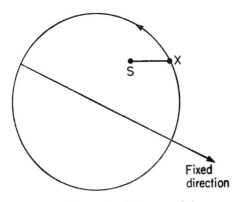

FIGURE 7. Hipparchus' theory of the seasons. As X moves around the circle, the segment drawn from X to the Sun S makes a constant angle with a fixed direction

spring-to-summer-to-autumn differs from autumn-to-winter-to-spring by three days. Indeed this fact was probably known long before the Greeks. It was explained by Hipparchus in the manner of Figure 7, where the point X moves uniformly in a circle around the Earth in a year. The Sun S is fastened to X by a short stick (not necessarily a real stick) which maintains a constant direction throughout the annual motion of X. Ugly and implausible or not, the Hipparchus theory grapples with the facts whereas the circular picture of Aristarchus fails already. It will already be clear then why it was so hard to arrive at the basic elements of a heliocentric theory. The two most striking bodies in the sky, the Sun and Moon, cause difficulties at the outset,

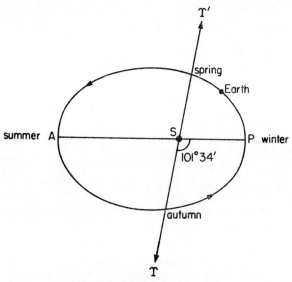

FIGURE 8. P is perihelion point, A is aphelion, and ♈ is the first point of Aries

even before we come to the planets. To make progress we have to be courageous enough (or foolhardy enough) to ignore the complexities of the Moon's orbit and also the known inequality of the seasons. As an aside, before we proceed, it may be of interest to compare the Hipparchus theory of Figure 7 with Figure 8, which gives the modern view of the inequality of the seasons.

If the reader now refers back to Plate 1 he may well wonder how a knowledge of the planetary motions can possibly make things any easier. Actually it does, as we shall see in the remainder of this chapter. First, there is the welcome simplification that the orbits of the planets are all more or less in the same plane as the path of the Sun. The measured tilts of the planes of the planetary orbits to the plane of the Sun's orbit for the six planets known to Copernicus and to the Greeks are given, together with other quantities of interest, in the following table:*

TABLE 3

Planet	Inclination to Sun's orbit	Period of revolution (*years*)	Longitude of perihelion	Distance
Mercury	7° 0′	0.2408	76° 13′	0.387
Venus	3° 24′	0.6152	130° 27′	0.723
Earth	–	1.0000	101° 34′	1.000 (standard)
Mars	1° 51′	1.8808	334° 35′	1.524
Jupiter	1° 18′	11.862	13° 2′	5.203
Saturn	2° 29′	29.457	91° 29′	9.539

The first step in breaking into the problem is to consider the observations of Venus. Venus sometimes runs ahead of the Sun in its path and sometimes it lags behind, in a kind of oscillatory motion, which we might try to represent in the manner of Figure 9, with Venus

* The gravitational influences of one planet on another cause their orbits to change slowly. The values in Table 3 apply at 1 January, 1920.

swinging in harmonic motion along a line through the Sun. In the part of the cycle from R to L, Venus would move more rapidly forward than the Sun, from L to R it would move less rapidly. At R the planet would be behind the Sun, at L it would be ahead. This picture therefore corresponds qualitatively with the observational situation. It fails, however, when we take account

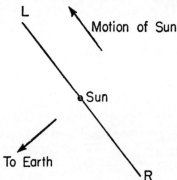

FIGURE 9. The apparent motion of Venus, as seen from the Earth, consists of an oscillation backwards from L to R, then forwards from R to L

of quantitative aspects of the data. On the model of Figure 9 we would expect variations in the brightness of Venus in the half-cycle from L to R to be exactly matched by similar variations in the opposite half-cycle from R to L. This is incorrect. The two parts L to R, R to L, are not the same, suggesting that the distance from the Earth to Venus is not the same. Nor indeed is the time of swing from L to R equal to the time from R to L, as it would be for simple harmonic motion.

A more sophisticated picture of the motion of Venus is shown in Figure 10. We now have a situation in which the Sun S moves uniformly in a circle around the Earth (ignoring the inequality of the seasons) and in which Venus V moves uniformly around the Sun in

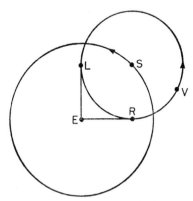

FIGURE 10. This model of the motion of Venus provided a first breakthrough in the understanding of the planetary motions. When projected on to the celestial sphere it gives the apparent oscillation of Figure 9

another circle. At point L of this second circle Venus is moving directly towards the Earth, at R Venus is moving away from the Earth. The situation projected on the sky is like Figure 9, but now the distance of Venus from the Earth is different over the sections L to R, R to L, and the times are also different. Also we have avoided the embarrassment that Figure 9 would require Venus to pass through the Sun.

The relative size relation of the two circles of Figure

10 is readily determined by measuring the angle between the direction of the Sun and the direction of Venus at the moment when Venus is most ahead of the Sun. A similar angle when Venus is most behind the Sun can also be measured, and the sum of these two angles is equal to the angle LER of Figure 10. With this known, the ratio of the radius SV of the moving circle to the radius ES is determined. It turns out to be about 0.723, the entry for Venus in the relative radius column of Table 3. The time for the motion L to R added to the time back again from R to L gives the period of Venus in its orbit about the Sun, 0.615 years.

We now have a model with predictive capacity. Although designed to fit a set of known data, it can be used to explain further new sets of data. This is a crucial requirement of a useful theory. Theories which only explain already known facts usually end up in a blind alley. The predictions of the model turn out to be quite good, for a reason we can understand from Table 1 of Chapter I. The eccentricity of the orbit of Venus happens to be very small, and the eccentricity of the Earth's orbit is reasonably small. So the actual orbits are nearly circles, which means that Figure 10 is a good approximation to the real situation. This regularity was the first to be discovered among the planets. It was known to Heraclides and was therefore available to Hipparchus and Ptolemy as well as to Aristarchus. From a code-cracking point of view, the case of Venus is a breakthrough.

The situation for Mercury is similar to that for Venus, but because of the large eccentricity of the orbit of Mercury the approximation given by a simple model like Figure 10 is not nearly so good. Fortunately, the shortcomings of a simple model were not so obvious in really ancient times, because reasonably accurate observations were not easy to make on account of the closeness of Mercury to the Sun. However, the discrepancies were well known to Ptolemy, who had considerable difficulty in improving on Figure 10.

The outer planets behave in a way that is both curiously similar and curiously different from the case of Venus. Imagine the Sun removed from Figure 10. We should not then be able to compare observationally the direction E to V with the direction of the Sun in the same way as before. We should be aware, however, that the arcs L to R, R to L were very different, because over the arc L to R the planet would appear to go backward on the sky, while over R to L it would appear to go forward. This alternation of backward and forward motions, with the forward motions winning on balance, is just what is observed for the outer planets. We look therefore to a model of the kind drawn in Figure 11, but with C now only a geometrical point, since the outer planets are not observed to oscillate about the position of the Sun. As in the case of Venus we seek to find the angle LER by comparing the lengths of the backward and forward motions. The matter is not so straightforward as before, however,

So far, the route towards understanding the motions of the planets is clearly marked—with the solar system being the way it is there is no other way of making a sensible start on the problem. But now we have alternative routes ahead. Which we elect to follow will depend on our temperament. If we are overwhelmingly concerned with the predictive quality of our model (which would be the point of view of most modern theoretical physicists) we shall seek to modify Figure 11 in order to secure a better correspondence between theory and observation. This is just what Ptolemy sought to do. But if we are concerned not so much with quantitative accuracy as with the qualitative structure of the problem, we shall be unable to overlook two further strange aspects of the situation.

It turns out that P in Figure 11 is required to move around the small circle in a year, the same period as S moves around E in Figure 10. But why should the Sun be involved with Figure 11 at all? Here we are concerned with the relation of the planet P, whether Mars, Jupiter, or Saturn, to the Earth. Indeed in Figure 11 we were required to *remove* the Sun from the centre of the small circle. Still more curious perhaps, if we ask where should the Sun be placed with respect to Figure 11 it turns out as in Figure 12, from observation it is found that a line drawn from the Earth in the direction of the Sun is always parallel to the line joining C to P.

Clearly this must mean *something*, but we cannot

decide *what* from observation alone. Indeed if observation had resolved this issue there would have been no controversy in astronomical history between geocentric and heliocentric theories. To distil meaning out of these curious facts we must take a theoretical step. Somewhere along the line EK of Figure 12 lies the Sun, but observa-

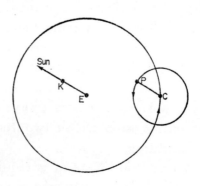

FIGURE 12. The Sun is observed to lie along a line through E drawn parallel to CP. The point K is chosen so that EK = CP

tion does not tell us where (at any rate observations available to the Greeks or to Copernicus). The inspired step taken by Aristarchus, and later independently by Copernicus, was to place the Sun at K, where K is chosen such that the length E to K is equal to the length C to P. It is worth redrawing Figure 12 with the Sun S placed firmly in this position, as in Figure 13, in order to draw out the geometrical fact that the motion of P described by Figure 13 is exactly the same as

Non docet instabiles Copernicus ætheris orbes,
Sed terræ instabiles arguit ille vices.

PLATE 4. Copernicus, based on self-portrait

PLATE 5. Heilsberg

PLATE 6. The Cathedral of Frauenburg

that described in Figure 14, where the planet P now goes in a larger circle about S.

It may take the reader a moment or two to become satisfied that Figures 13 and 14 do indeed give the same

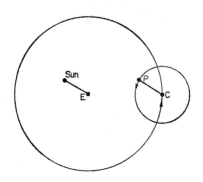

FIGURE 13. This is the same diagram as Figure 12 but with the Sun placed by hypothesis at the point K. This was the crucial step needed to arrive at a heliocentric theory

motion of P. The equivalence of these two pictures was known already to Apollonius who lived in the third century B.C., long before Ptolemy (*ca* A.D. 150). The reader familiar with complex numbers will note that if E is taken as origin and z_1, z_2 are complex numbers determining the points S, C respectively, then P in Figure 14 is $z_1 + z_2$, while P in Figure 13 is $z_2 + z_1$. Because complex numbers are commutative the positions of P are the same in the two cases. In the next chapter we shall find that the constructions discovered by

Ptolemy and Copernicus can be understood in an elegant way by using complex numbers.

We have a construction of the form of Figure 10 for an inner planet and one of the form Figure 14 for an outer

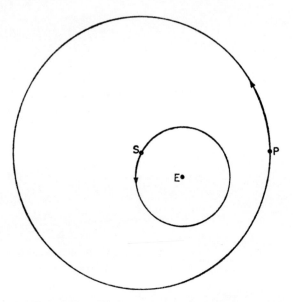

FIGURE 14. The planet P now moves in a circle about S, while S moves in a circle about E. The motion of P is exactly the same as in Figure 12

planet. Drawn together in Figure 15, we can at last take advantage of the fact that the motion of S around E is kinematically equivalent to the motion of E around S. Thus Figure 15 is equivalent to Figure 16. The respective circles have the same radii in the two figures, so

that by using the same periods of motion in corresponding circles the relative positions of E, S, V, P
must be the same in the two pictures at all times. Here
P can represent any one of the outer planets and V can

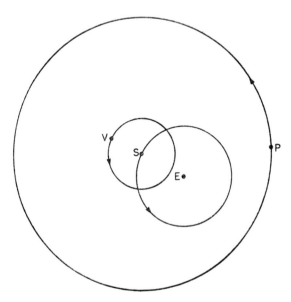

FIGURE 15. This combines Figures 10 and
14, with V representing an inner planet and P
representing an outer planet

represent either Venus or Mercury. For the six planets
observed in ancient times, using the relative radii of
Table 3, we get the structure shown in Figure 17. This
is the heliocentric theory of Aristarchus.

Copernicus probably rediscovered the line of reasoning
set out above before returning from Italy to Poland in

1503. I suggested in the previous chapter that the year 1513 may well have represented a watershed both in his life and in the development of the heliocentric theory.

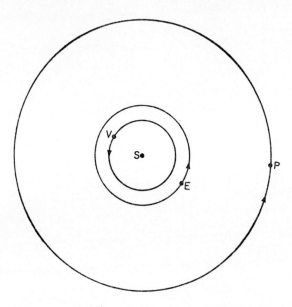

FIGURE 16. The motions in this heliocentric picture are exactly the same as in Figure 15, because the motion of S about E in Figure 15 is kinematically equivalent to the motion of E about S in this figure

The tendency to suppose that Copernicus discovered his theory early in life arises, I believe, from a misjudgment of the magnitude of his achievement. The popular view, and I think to some extent the view of the less technical biographies, is that the simple helio-

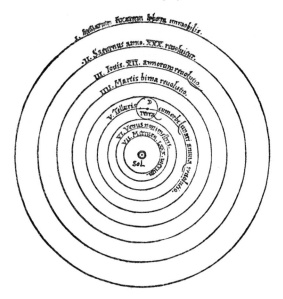

FIGURE 17. The heliocentric picture for the
planets known in the time of Copernicus, as
it was drawn in *de revolutionibus*

centric theory described above represents the major
part of what he did. So far from this being the case, we
have as yet covered only the easy beginning. For
Copernicus, the hard part, the fight with Ptolemy, had
still to come.

IV
Copernicus and Ptolemy

BOTH COPERNICUS AND PTOLEMY ARRIVED AT their constructions from the analysis of a vast amount of data. Simply to describe the constructions here without including explicit comparison with observation would be a somewhat pointless procedure. We should be involved in geometrical intricacies whose *raison d'être* was unclear. The need to include long and extensive reference to the data can be avoided, however, if we start from the elliptic orbit given in equation (6) of Chapter I, for we know that such an orbit represents all the data correctly (at any rate to within the margins we are interested in here). So we can convert the question of how the constructions of Ptolemy and Copernicus compare with observation to the question of how the constructions compare with the elliptic orbit. This is the approach to be followed in the present chapter. Instead of following historic evolution, in which data are used in a struggle towards a more correct theory, we start with the correct theory and see to what extent it has to be degraded in order to arrive at the methods used by Ptolemy and Copernicus.

The starting point, then, is the equation

$$r = \frac{a(1 - e^2)}{1 + e \cos \theta} \qquad (7)$$

in which r is distance measured from the Sun, e is the eccentricity of the ellipse, and θ represents the angle

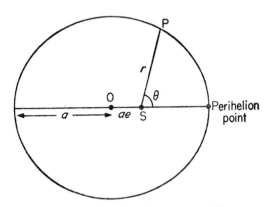

FIGURE 18. The angle θ is measured from the perihelion point. The distance from the centre O to the focus S is ae, and the distance from S to P is r

between the major axis and the line drawn from the Sun to the planet. The relation of the focus to the centre of the ellipse is shown in Figure 18, the distance being the product ae. The distance of the planet from the Sun at perihelion is $a(1 - e)$ and at aphelion $a(1 + e)$.

We also need Kepler's law of equal areas. Suppose in a small time interval dt the line from S to P turns through an element of angle $d\theta$. If r were to remain unchanged, the line from the Sun to the planet would sweep out a narrow isosceles triangle of height r and of small base $r\,\mathrm{d}\theta$, and so would have area $\frac{1}{2}r^2\,\mathrm{d}\theta$. Even if r changes a little in the time $\mathrm{d}t$ the area remains $\frac{1}{2}r^2\,\mathrm{d}\theta$ to within the first order of smallness. The rate at which the planet sweeps out an area is therefore $\frac{1}{2}r^2\,\mathrm{d}\theta/\mathrm{d}t$, any small term due to changing r becoming negligibly small in the limit as $\mathrm{d}t$ tends to zero. It follows from Kepler's law that $r^2\,\mathrm{d}\theta/\mathrm{d}t$ is a constant,

$$r^2\frac{\mathrm{d}\theta}{\mathrm{d}t} = h \qquad (8)$$

where h is a constant. The mean motion n was defined in equation (3) of Chapter I. According to this equation the time required for a planet to make a complete revolution in its orbit is $2\pi/n$. In this length of time the total area swept out by the line from S to P is

$$\frac{1}{2}r^2\frac{\mathrm{d}\theta}{\mathrm{d}t} \cdot \frac{2\pi}{n} = \pi\frac{h}{n} \qquad (9)$$

which must be equal to the area πab of the whole ellipse. Using equation (1) of Chapter I we can write πab in the form $\pi a^2(1 - e^2)^{\frac{1}{2}}$, and this is equal to equation (9),

$$\pi a^2(1 - e^2)^{\frac{1}{2}} = \pi\frac{h}{n}, \qquad (10)$$

or simply

$$n = \frac{h}{a^2(1 - e^2)^{\frac{1}{2}}} \qquad (11)$$

for the relation of the mean motion to h, a, e.

Eliminating r between equations (7) and (8) gives

$$\frac{\mathrm{d}\theta}{\mathrm{d}t} = \frac{h}{a^2(1 - e^2)^2}[1 + e\cos\theta]^2 = \frac{n}{(1 - e^2)^{\frac{3}{2}}}[1 + e\cos\theta]^2.$$

$$(12)$$

If we could conveniently integrate this equation to determine θ in terms of the time t, obtaining some explicit function, $\theta(t)$ say, we should have a closed solution to the problem of determining the motion with respect to time. $\theta(t)$ would give the heliocentric longitude at time t, while the substitution of $\theta(t)$ in equation (7) would give the heliocentric distance. This cannot be done, however. Equation (12) can only be integrated as an infinite series, indeed the series given already in (2) of Chapter I,

$$\theta = nt + 2e \sin nt + \tfrac{5}{4}e^2 \sin 2nt + \cdots. \qquad (13)$$

Substitution of (13) in (7) gives the following infinite series for r,

$$\frac{r}{a} = 1 - e \cos nt + \tfrac{1}{2}e^2(1 - \cos 2nt) + \cdots. \qquad (14)$$

We already remarked in Chapter I that it is this feature of the correct solution which made the problem of determining the planetary motions so awkward.

Suppose we represent the planet by a complex number

$$z = r \exp i\theta = a(1 - e \cos nt + \cdots)$$
$$\exp i(nt + 2e \sin nt + \cdots). \qquad (15)$$

Now degrade the problem by working only to the first order in the eccentricity e. It is then sufficient to write

$$\exp(2ie \sin nt) \doteq 1 + 2ie \sin nt \qquad (16)$$

and for (15) we have

$$z \doteq a(1 - e \cos nt + 2ie \sin nt) \exp int$$
$$= a(1 + \tfrac{1}{2}e \exp int - \tfrac{3}{2}e \exp -int) \exp int$$
$$= -\tfrac{3}{2}ae + a \exp int + \tfrac{1}{2}ae \exp 2int. \qquad (17)$$

Evidently the position P of the planet can be represented in this order of approximation by the following rules:

(1) make a displacement of length $\frac{3}{2}ae$ from S to a point K in a negative sense along the real axis,

(2) then make a displacement a exp int from K to a point L, and

(3) finally make a displacement $\frac{1}{2}ae$ exp $2int$ from L to P.

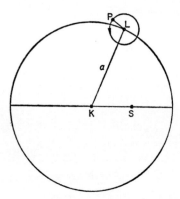

FIGURE 19. This is the construction of Copernicus. The line KL turns at an angular rate equal to the mean motion n of the planet, whereas LP turns at an angular rate $2n$. The length LP is $\frac{1}{3}$ of KS, which is $\frac{3}{2}ae$

These rules are shown geometrically in Figure 19. The point K stays fixed. As t changes, the point L moves with angular velocity n about K in a circle of radius a, the sense of the motion being anticlockwise, while the planet P itself moves about L with angular velocity $2n$ in a circle of radius $\frac{1}{2}ae$, also in an anticlockwise sense.

The angular velocities here are measured of course with respect to a fixed direction—the real axis of our complex numbers. This is the construction of Copernicus.

Actually, Copernicus made two minor deviations from these rules, deviations which were not improvements, and which were subsequently corrected by Kepler in his earliest attempts on the problem. The small circle from L to P was omitted in the case of the Earth. Copernicus also replaced the Sun S, in all planetary constructions other than that of the Earth, by the point K determined for the Earth. This meant that the Copernican theory of the planetary motions was not strictly heliocentric to the first order in the eccentricity of the Earth's orbit.

It has to be remembered, however, that Copernicus felt called on to face up to every problem contained in the data available to him. In the previous chapter we saw how he became concerned with the difficult problem of the motion of the Moon. It is worth interrupting the present discussion for a moment to give a second example.

The planetary orbits are not strictly ellipses, as we have so far taken them to be, because one planet disturbs the orbit of another through the gravitational force which it exerts. For the most part these effects are slight, but the fact that the line S to K of Figure 19 slowly changes for the Earth's orbit was known to Copernicus, from a comparison of the observations of Arabic astronomers with those of the Greeks. As always,

Copernicus felt obliged to include even this minor effect in his kinematical description. From certain other comparisons it also seemed to him as if the Earth's whole orbit were subject to oscillation, 'trepidations' as he called them. These effects, later shown by Tycho Brahe to be illusory, did not survive to trouble Kepler. But they were a serious worry to Copernicus, who found himself trying to represent in as simple a way as possible, a considerable catalogue of small motions, some of them merely artifacts of bad data. In retrospect, it is of course easy for us to see that Copernicus should have been less punctilious. He would have done better simply to have applied the construction of Figure 19 consistently to all planets—but this is to be wise after the event.

Returning now to our main argument, equation (17) can be rewritten in the form

$$z \doteq a(1 - e \cos nt + 2ie \sin nt) \exp int$$
$$= a(1 + e \cos nt) \exp int - 2ae(\cos nt - i \sin nt) \exp int$$
$$= a(1 + e \cos nt) \exp int - 2ae. \qquad (18)$$

Hence we can also arrive at the position of a planet, again correct to the first order in the eccentricity, by using the following alternative rules:

(1) make a displacement of length $2ae$ from S to a point A in the negative sense along the real axis and

(2) make a displacement of length $a(1 + e \cos nt)$

from A in a direction making an angle *nt* with the positive real axis.

How shall we interpret this construction with respect to changing time? Turning to Figure 20 we see that the position P for the planet lies, to the first order in the

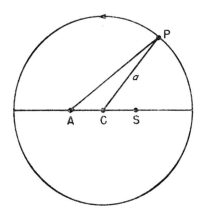

FIGURE 20. This is the basic construction of Ptolemy, but applied in a heliocentric picture. Here AC = CS = *ae* (these distances are exaggerated) and the line from A to P turns at an angular rate equal to the mean motion *n*

eccentricity, on the circle of radius *a* with centre C at the mid-point of AS. This is because the distance from A to P is $a(1 + e \cos nt)$, the angle PAS being *nt*. This alternative way of representing the motion of the planet uses only one circle. The odd feature is that the radius C to P does not turn at a uniform rate. It is the line A to P which turns uniformly. Copernicus was aware that the motion could be represented this way but he disliked the concept of turning about the centre of a circle at a non-uniform rate. Kepler had no such scruples. He seems to have preferred Figure 20 to Figure 19.

Except that it is related to motion about the Sun instead of about the Earth, Figure 20 is the construction of Ptolemy. The point A in his *punctum aequans*. Write the position of the Earth relative to the Sun in the form

$$z_E \doteq a_E[1 + e_E \cos n_E (t - t_E)] \exp i n_E(t - t_E) - 2a_E e_E,$$

(19)

and write a similar equation for a planet P,

$$z_P \doteq a_P[1 + e_P \cos n_P t] \exp i n_P t - 2a_P e_P.$$ (20)

Notice that from the beginning, in equation (13), we have taken $\theta = 0$ at $t = 0$. This meant that throughout the above discussion of the motion of a single planet we measured the time t from a moment when the planet was at perihelion. Now that we are considering two planets, E and P, we cannot suppose that both were simultaneously at perihelion. According to equation (19), E is at perihelion at $t = t_E$, and according to equation (20), P is at perihelion at $t = 0$.

The position of P relative to E is given by the simple subtraction of (19) from (20),

$$z_P - z_E \doteq a_P[1 + e_P \cos n_P t] \exp i n_P t - 2a_P e_P -$$
$$a_E[1 + e_E \cos n_E(t - t_E)] \exp i n_E(t - t_E) + 2a_E e_E.$$ (21)

The form of equation (21) is not as straightforward to handle geometrically as equation (18), and to overcome the added complexity occasioned by this subtraction Ptolemy was forced to degrade his theory slightly by omitting further small terms. He would have done better by working with equation (18), i.e. by using a heliocentric theory. Reference back to Table 1 of Chapter I shows that the eccentricities of all planetary orbits but that of Venus are large compared to the eccentricity e_E of the Earth's orbit. Suppose for all planets except Venus we omit terms involving e_E from (21),

$$z_P - z_E \doteq a_P\{(1 + e_P \cos n_P t) \exp i n_P t - 2e_P\} - a_E \exp i n_E(t - t_E). \quad (22)$$

We can represent the first term here by exactly the same construction as that of Figure 20. So to obtain P relative to E, according to equation (22), we only have to add to Figure 20 the effect of the second term. Noting that

$$- a_E \exp i n_E(t - t_E) = a_E \exp i n_E(t + \pi - t_E), \quad (23)$$

the required construction has the form given in Figure 21. We now have two circles, as in the Copernicus construction of Figure 19. The point L moves about the circle of centre C and radius a_P, but with constant angular velocity n_P about the *punctum aequans* A, and the planet P moves around a circle of radius a_E and centre L with angular velocity n_E. The position of P on this circle has phase $\pi - t_E$ at $t = 0$.

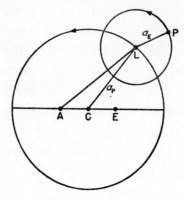

FIGURE 21. The construction of Ptolemy
applied to a geocentric picture. The line AL
turns with the mean motion n_P of the planet,
while the line LP turns with the mean motion
n_E of the Earth

To use the Ptolemaic construction of Figure 21 one
needs in practice to determine the following quantities
for each planet:

the ratio a_P/a_E,
the eccentricity e_P,

a time when the planet is at perihelion. This is the same
data as is needed to make practical application of the
Copernicus construction of Figure 19. The a_P/a_E values
used by Ptolemy and Copernicus are given in the
table on facing page.

There is no entry in the first column for the case of the
Earth because the motion of the Sun in the Ptolemaic
theory is represented by $-z_E$ and z_E, given by equation

(19), is represented by the construction of Figure 20 in which there is only one circle.

In all cases in Table 4 the difference between the modern value and the value used by Copernicus is in the same sense as the value used by Ptolemy, which indicates the extent to which Copernicus relied on Ptolemy's data. Copernicus evidently used Ptolemy's values as a working basis, modifying them somewhat to accord with his own observations and with those of Arabic astronomers. The situation seems to have been the same for the e_P values used in the two theories.

TABLE 4

Planet	a_P/a_E Ptolemy	a_P/a_E Copernicus	a_P/a_E Modern
Mercury	0.3708	0.3763	0.3871
Venus	0.7194	0.7193	0.7233
Earth	—	1.0000	1.0000
Mars	1.5191	1.5198	1.5237
Jupiter	5.2164	5.2192	5.2028
Saturn	9.2336	9.1743	9.5388

From what has so far been said it will be realized that the predictive quality of the constructions of Ptolemy and Copernicus are very nearly the same. Indeed the error in the Copernican theory occasioned by taking the point K determined for the Earth's orbit as the point S for the other planets meant that the Copernican theory involved just the same approximation as we made in passing from equation (21) to equation (22). Terms of order $a_E e_E$ were neglected in both theories. The

Copernican theory becomes superior to the Ptolemaic theory, however, when account is taken of the inclinations of the planetary orbits (see Table 3 of Chapter III). Copernicus took all the planes of the planetary orbits to pass through the point K for the Earth's orbit. Since K lies close to the Sun this was nearly the same as taking the Sun to lie in the plane of every orbit, which is the correct situation. Ptolemy, on the other hand, took all the planes to pass through the Earth, which is a worse approximation. Consequently the Copernican theory of the motions of the planets in latitude is better than the Ptolemaic theory.

I have not sought in the above discussion to deviate from standard accounts of the two theories (for example, J. L. E. Dreyer, *A History of Astronomy from Thales to Kepler*, Dover Publications, 1953). I have simply interpreted these discussions in a modern sense through working directly from the elliptic orbits of the planets. At this stage I wish to take issue, however, with a criticism which is sometimes made of Ptolemy. It is said that Ptolemy's construction for Venus is a weakness of the theory, since it differs from Figure 21. But the change of construction for Venus shows that Ptolemy knew better than his critics what he was doing. It will be recalled that we passed from equation (21) to equation (22) on the basis that e_E, the eccentricity of the Earth's orbit, is smaller than e_P, the eccentricity of the planet's orbit. We remarked that this is true for every planet except Venus. To obtain a corresponding

construction for Venus we must therefore work by
omitting the e_P terms in equation (21), writing

$$z_P - z_E \doteq -a_E\{[1 + e_E \cos n_E(t - t_E)] \exp in_E(t - t_E)$$
$$- 2e_E\} + a_P \exp in_Pt \quad (24)$$

in place of equation (22). The construction for equation
(24) is like Figure 21 but with $EC = AC = a_E e_E$
instead of $a_P e_P$, with a_E for CL instead of a_P, and with
a_P for LP instead of a_E. The construction has become
inverted with respect to E and P.

This ends our interpretation of the work of Ptolemy
and Copernicus. It seems truly remarkable in retrospect
that Ptolemy did not realize that the second term in
equation (22), which is the term giving rise to the
epicyclic motion of P about L in Figure 21, could be
interpreted as the Earth's motion reflected in the motions
of the planets. This is clearly seen in the above treat-
ment, for this particular term appeared when we sub-
tracted the position of the Earth z_E from the position
of the planet z_P. It would be valuable, as a matter of
historical interest, to go over Ptolemy's work (in as
near to its original form as possible) to find if there are
any hints as to what his position really was. Did he
really overlook the source of the epicycle in Figure 21?
Or was he so overwhelmingly concerned with the
predictive quality of his theory that the issue did not
concern him very greatly? I am led to wonder how far
editors of the *Almagest*, like the editor of *de revolutionibus*
may have inserted their own interpretations and preju-
dices into Ptolemy's work at just this point.

Epilogue from the Twentieth Century

AT THE BEGINNING OF CHAPTER I IT WAS STATED that we can take either the Earth or the Sun, or any other point for that matter, as the centre of the solar system. This is certainly so for the purely kinematical problem of describing the planetary motions. It is also possible to take any point as the centre even in dynamics, although a recognition of this freedom of choice had to await the present century. Scientists of the nineteenth century felt the heliocentric theory to be established when they determined the first stellar parallaxes. The positions of nearby stars were found to undergo annual oscillations which were taken as reflections of the Earth's annual motion around the Sun. But, kinematically speaking, we can always give to the stars epicyclic motions similar to the ones we found in Chapter IV for the planets. Indeed, if we wish to consider the Earth to be at rest, it will be necessary to give an annual epicyclic motion to every object in the distant universe, as well as to the planets of the solar system. We cannot dismiss such a procedure simply on the grounds of inconvenience or absurdity. If our feeling that the Earth really goes around the Sun, not the Sun around the Earth, has any objective validity there must be some important physical property, expressible in precise mathematical terms, which emerges in the heliocentric picture but not in a geocentric one. What can this property be?

Consider the well-known Newtonian equation

$$\text{mass} \times \text{acceleration} = \text{force}.$$

The mass which appears here for a particular body is intended to be always the same, independent of where the body is situated or of how it is moving. Suppose we describe the position of a body as a function of time in some given reference frame, and suppose we know the mass. Then provided we also have explicit knowledge of the force acting on the body, Newton's equation gives us the acceleration. Determining the motion is from there on a mathematical problem only—in technical terms we have to integrate the above equation. This procedure, which forms the basis of Newtonian mechanics, fails unless we know the force explicitly. In the Newtonian theory of the planetary motions, the theory leading to the basic ellipse from which we worked in Chapter IV, the force is taken to be given by the well-known inverse square law:

Two masses m_1, m_2, distant r apart, attract each other with a force

$$G\frac{m_1 m_2}{r^2}$$

where *G* is a numerical constant. The force is directed along the line joining the bodies.

Now comes the critical question:

In what frame of reference is this law considered to operate?

In the solar system we cannot consider the inverse square law to operate *both* in the situation in which the Sun is taken as the centre and in which the Earth is taken as centre, because Newton's equation would then lead to contradictory results. We should find a planet following a different orbit according to which centre we chose, and a body cannot follow two paths (at any rate not in classical physics). It follows that in order to use the inverse square law in a constructive way we must make a definite choice of centre. The situation which now turns out is that to obtain results agreeing with observation it is the Sun which must be chosen as centre. If the Earth were chosen instead, some law of force other than the inverse square law would be needed to give motions agreeing with observation.

Although this argument was believed in the nineteenth century to be a satisfactory justification of the heliocentric theory, there were causes for disquiet if one looked into it a little more carefully. When we seek to improve the accuracy of calculation by including mutual gravitational interactions between planets we find— again in order to calculate correctly—that the centre of the solar system must be placed at an abstract point known as the 'centre of mass' which is displaced quite appreciably from the centre of the Sun. And if we imagine a star to pass moderately close to the solar system, in order to calculate the perturbing effect correctly, again using the inverse square rule, it would be essential to use a 'centre of mass' which included the star. The 'centre' in this case could lie even farther away from the centre of the Sun. It appears then that the 'centre' to be used for any set of bodies depends on the way in which the local system is considered to be isolated from the universe as a whole. If a new body is added from outside to the set, or if a body is taken away, the 'centre' changes.

A similar circumstance was already present throughout our calculations, when we regarded angles as being measured with respect to a 'fixed direction', it being implied that distant stars had directions that were 'fixed' in this sense. If we make a calculation, using both Newton's equation and the inverse square law, but measuring angles with respect to a direction that rotates with respect to the distant universe, things go very badly wrong. Newton was fully aware that his system of dynamics would only work correctly provided the 'fixed directions' in the theory were chosen in a suitable way. His reference to the well-known rotating bucket experiment was intended to illustrate this point.

It is clear therefore that in order to define the appropriate 'centre' of the local system in a useful way, and in order to define 'fixed directions' relative to which angles are to be measured, we must take account of the relation of the local system to the universe outside. It seems that the local laws of force only take simple forms when the centre is unaccelerated with respect to a frame of reference determined by the universe in the large, and when the fixed directions do not rotate with respect to the distant universe. From this point of view we can compare the heliocentric and geocentric theories of the solar system in an unequivocal way. We ask:

Is it the Sun that is accelerated with respect to the universe, or is it the Earth?

Neglecting small effects, the answer is that the Earth is accelerated, not the Sun. Hence we must use the helio-centric theory if we wish to take advantage of simple rules for the local forces. But this is not to say that we cannot use the geocentric theory if we are willing to use more complex rules for the forces.

The present discussion has been formulated from the standpoint of the Newtonian theory, which is not well suited to problems concerning the universe in the large. We might hope therefore that the Einstein theory, which is well suited to such problems, would throw more light on the matter. But instead of adding further support to the heliocentric picture of the planetary motions the Einstein theory goes in the opposite direction, giving increased respectability to the geocentric picture. The relation of the two pictures is reduced to a mere coordinate transformation and it is the main tenet of the Einstein theory that any two ways of looking at the world which are related to each other by a coordinate transformation are entirely equivalent from a physical point of view. Moreover, the method of calculating the effect of gravitation is changed in the Einstein theory to a form which applies equally to all such related ways of expressing a problem.

It may still happen that it is easier to work through the details of a particular problem with respect to one coordinate system rather than another, but no special physical merit is to be adduced from such a circumstance. Indeed, from a mathematical point of view, the problem of the planetary motions certainly continues to be easier to grapple with in the heliocentric picture. The simplification of such a picture shows itself in the Einstein theory through boundary conditions which are impressed on the spacetime structure at a large distance from the Sun—which is to say in terms of the control impressed by the universe in the large.

So we come back full circle to what was said at the beginning of this book. Today we cannot say that the Copernican theory is 'right' and the Ptolemaic theory 'wrong' in any meaningful physical sense. The two theories, when improved by adding terms involving the square and higher powers of the eccentricities of the planetary orbits, are physically equivalent to one another. What we can say, however, is that we would

Bibliography

ARMITAGE, A., *The World of Copernicus* (New American Library, 1951).

COPERNICUS, N., *Original Manuscript* (Basingstoke: Macmillan, 1972).

DREYER, J. L. E., *A History of Astronomy from Thales to Kepler* (London: Dover Publications, 1953).

MIZWA, S. P., *Nicolas Copernicus* (Kosciuszko Foundation, 1943).

PROWE, L., *Nicolaus Coppernicus* (Berlin, 1883–4).

ROSEN, E., *Three Copernican Treatises* (London: Dover Publications, 1959).

RYBKA, E., *Four Hundred Years of the Copernican Heritage* (Cracow, 1964).

Index